FROZEN VEGETABLES

by Gretchen Will Mayo

Reading consultant: Susan Nations, M.Ed., author/literacy coach/consultant

WEEKLY WR READER®
EARLY LEARNING LIBRARY

Please visit our web site at: **www.garethstevens.com**
For a free color catalog describing Weekly Reader® Early Learning Library's
list of high-quality books, call 1-877-445-5824 (USA) or 1-800-387-3178 (Canada).
Weekly Reader® Early Learning Library's fax: (414) 336-0164.

Library of Congress Cataloging-in-Publication Data

Mayo, Gretchen.
    Frozen vegetables / by Gretchen Will Mayo.
      p. cm. — (Where does our food come from?)
      Summary: Discusses vegetables and explains how they are frozen so that they will not
spoil as quickly.
    Includes bibliographical references and index.
    ISBN 0-8368-4066-6 (lib. bdg.)
    ISBN 0-8368-4073-9 (softcover)
    1. Frozen vegetables—Juvenile literature. [1. Vegetables.] I. Title.
TP372.3.M39   2004
664'.853—dc22
                                     2003061009

This edition first published in 2004 by
Weekly Reader® Early Learning Library
A Member of the WRC Media Family of Companies
330 West Olive Street, Suite 100
Milwaukee, WI 53212 USA

Editor: JoAnn Early Macken
Art direction, cover and layout design: Tammy West
Photo research: Diane Laska-Swanke

Photo credits: Cover (main), title, pp. 4, 5, 6, 7, 8, 9, 10, 11, 12, 13, 14, 15, 16, 17, 18, 19, 20 © Gregg Andersen;
cover (background) © Diane Laska-Swanke

Printed in the United States of America

CPSIA Compliance Information Batch # CR211220GS: For further information contact Gareth Stevens, New York, New York at 1-800-542-2596.

Table of Contents

A family gets ready to can vegetables.

Canning and Freezing

Vegetables are good for us. We should each eat
three servings a day. Before people had freezers
in their homes, that was hard to do. Long ago,
people had to store, or can, their own vegetables
in jars. That was the only way to eat vegetables
in the winter. They did not taste fresh.

Bacteria could spoil vegetables that were canned at home. Freezing stops bacteria from spoiling food. A bag of frozen vegetables can keep for years. You can buy fresh green beans from a farm or a store in summer. Pop them in the freezer. They will still taste fresh at Thanksgiving dinner.

Frozen vegetables add vitamins to the menu.

Growing vegetables line up on a
truck farm.

Growing Vegetables

Today, we have many vegetable choices.
Store freezers are full of frozen vegetables.
Most vegetables grow on special farms called
truck farms.

Truck farmers grow large fields full of vegetables. Truck farmers sell the vegetables they grow. Groceries or food companies buy them. Some ripe vegetables are trucked to a factory.

A farmer loads onions into a shipping carton.

Bad weather can ruin a vegetable crop.

Good weather is the truck farmer's friend. Farmers need good weather for planting, growing, and harvesting. During planting season, the fields cannot be too wet. Growing plants need rain, but not too much rain. They need sun, but not too much sun.

Frozen food companies look for the ripest vegetables. They buy only the best. Corn kernels cannot be too small or too large. Green beans must be crisp yet tender. Carrots with cracks are not good enough.

A truck farmer's vegetables are inspected carefully.

Machines clean dirt from ears of corn.

The Frozen Food Factory

Fresh vegetables are sent to the frozen food company, or plant. In most plants, people do not touch the vegetables. Machines clean them.

Machines with knives snip the ends off of green beans. Machines pull husks off of corn. Machines take the shells off of peas.

Ears of corn are stripped by machines.

Good ears of corn are pushed
onto the moving belt.

A belt carries the vegetables through the plant.
A camera looks at each vegetable. It looks for
flaws or damage. A blast of air kicks the bad
vegetables out of line.

The good vegetables move on. Machines slice green beans. Broccoli is chopped. Corn kernels are cut from the cob. A shaker tub separates the small and large peas.

Carrots are cleaned, sliced, and frozen in the plant.

A worker looks on the belt for bad corn.

Stray corn silk and husk pieces are removed. Green bean ends and pieces of pea shells are whisked away. Workers inspect the vegetables again.

The factory can freeze food faster than we can at home. Vegetables hold a large amount of water. Water expands when it is frozen.

A blast of icy air freezes corn kernels.

The soft zucchini on the left was frozen at home.
The zucchini on the right was rapidly frozen.

Home freezing turns the water into chunks of ice.
When the vegetable is thawed, it turns soft. Rapid
freezing at the factory turns the water into tiny
crystals. Vegetables stay firmer when they thaw.

The food factory can freeze each vegetable piece alone. A machine blows freezing air over the pieces. At once, they freeze solid. Rapid freezing keeps them from sticking together. The pieces are stored in cardboard boxes called totes. Each tote can hold 1,200 pounds (544 kilograms) of vegetables.

Frozen kernels of corn are dumped into a tote.

Workers stack sealed packages
of corn on the cob.

The totes are stored in big freezer buildings.
Later, workers return the totes to the factory. The
frozen vegetables are packaged there. Machines
seal vegetables in packets. The packets are placed
in cartons.

The cartons are stored at the factory. Then the frozen vegetables are sent to stores, hospitals, and schools.

Cartons wait in cold storage rooms.

Many frozen vegetables taste delicious in soup.

Vegetable Choices

Lettuce and cabbage do not freeze well. Celery and tomatoes do not freeze well. They turn to mush when they thaw. Cabbage, celery, and tomatoes can be frozen in soups or sauces.

Vegetables are an important part of good eating habits. They are full of vitamins and minerals. They help keep us healthy and make us strong. Frozen vegetables let us have choices any time.

The food pyramid is our guide to healthful eating.

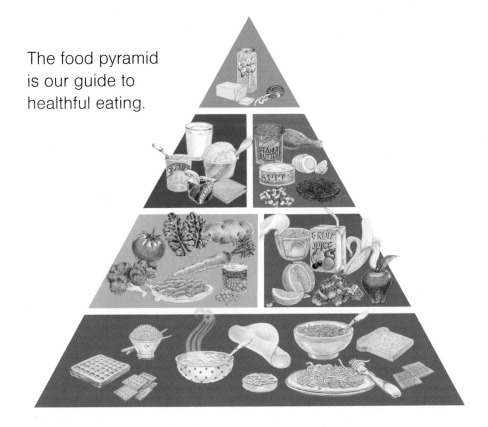

Glossary

bacteria — tiny living organisms that can cause disease

cartons — containers used to store or ship things

crystals — solid pieces of a substance that have flat surfaces joined at sharp angles

expands — becomes larger

husks — dry or leaflike outer coverings

kernels — grains on a corn cob

mush — a thick, soft mass

tender — not tough

truck farms — farms that raise vegetables that are usually trucked to stores or factories

For More Information

Books

Aliki. Corn Is Maize: *The Gift of the Indians*. NY: HarperCollins, 1986.

Ehlert, Lois. *Growing Vegetable Soup*. Orlando, FL: Harcourt Brace Jovanovich, 1987.

Maurer, Tracy. *Growing Vegetables. Green Thumb Guides*. Vero Beach, FL: Rourke Book Co., 2001.

Nelson, Robert. *Vegetables. Food Groups* Series. Minneapolis: Lerner, 2003.

Web Sites

The Dining Room Game
www.welltown.gov.uk/school/dining_game.html
Test your food-group knowledge

Have a Bite Cafe
exhibits.pacsci.org/nutrition/cafe/cafe.html
Build a meal and get nutrition information

Meals Matter Activities—My Very Own Pizza
www.mealsmatter.org/Activities/mvop.htm
Build your own pizza

Index

About the Author

Gretchen Will Mayo likes to be creative with her favorite foods. In her kitchen, broccoli and corn are mixed with oranges to make a salad. She sprinkles granola on applesauce. She blends yogurt with orange juice and bananas. She experiments with different pasta sauces. When she isn't eating, Ms. Mayo writes stories and books for young people like you. She is also a teacher and illustrator. She lives in Wisconsin with her husband, Tom, who makes delicious soups. They have three adult daughters.